United States Government Accountability Office

Report to Congressional Requesters

April 2012

INFORMATION TECHNOLOGY REFORM

Progress Made; More Needs to Be Done to Complete Actions and Measure Results

GAO-12-461

GAO Highlights

Highlights of GAO-12-461, a report to congressional requesters

April 2012

INFORMATION TECHNOLOGY REFORM

Progress Made; More Needs to Be Done to Complete Actions and Measure Results

Why GAO Did This Study

While investments in IT have the potential to improve lives and organizations, federal IT projects too often experience cost overruns, schedule slippages, and performance shortfalls. To address acquisition challenges, improve operational efficiencies, and deliver more value to the American taxpayer, in December 2010, OMB's Federal CIO issued a 25-point IT Reform Plan.

GAO was asked to (1) evaluate the progress OMB and key federal agencies have made on selected action items in the IT Reform Plan, (2) assess the plans for addressing action items that are behind schedule, and (3) assess the extent to which sound measures are in place to evaluate the success of the IT reform initiatives. To do so, GAO selected 10 of the 25 action items from the IT Reform Plan, focusing on the more important activities due to be completed by December 2011; analyzed agency documentation; and interviewed agency officials.

What GAO Recommends

GAO is making recommendations to three agencies to complete key IT Reform action items; the agencies generally concurred. GAO is also making recommendations to OMB to complete key action items, accurately characterize the items' status, and establish measures for IT reform initiatives. OMB agreed to complete key action items, but disagreed with the latter recommendations, noting that the agency believes it is characterizing the items' status correctly and that measures are not warranted. GAO maintains that its recommendations are valid.

View GAO-12-461. For more information, contact David A. Powner at (202) 512-9286 or pownerd@gao.gov.

What GAO Found

The Office of Management and Budget (OMB) and key federal agencies have made progress on action items in the Information Technology (IT) Reform Plan, but there are several areas where more remains to be done. Of the 10 key action items GAO reviewed, 3 were completed and 7 were partially completed by December 2011, in part because the initiatives are complex (see table). OMB reported greater progress than GAO determined, stating that 7 of the 10 action items were completed and that 3 were partially completed. While OMB officials acknowledge that there is more to do in each of the topic areas, they consider the key action items to be completed because the IT Reform Plan has served its purpose as a catalyst for a set of broader initiatives. They explained that work will continue on all of the initiatives even after OMB declares that the related action items are completed under the IT Reform Plan. We disagree with this approach. In prematurely declaring the action items to be completed, OMB risks losing momentum on the progress it has made to date. Until OMB and the agencies complete the action items, the benefits of the reform initiatives—including increased operational efficiencies and more effective management of large-scale IT programs—will likely be delayed.

OMB and key agencies plan to continue efforts to address the seven items that GAO identified as behind schedule, but lack time frames for completing most of them. For example, OMB plans to work with congressional committees during the fiscal year 2013 budget process to assist in exploring legislative proposals to establish flexible budget models and to consolidate certain routine IT purchases under agency chief information officers (CIO). However, OMB has not established time frames for completing five of the seven IT Reform Plan action items that are behind schedule. Until OMB and the agencies establish time frames for completing these corrective actions, they increase the risk that key action items will not be completed or effectively managed to closure. Further, they diminish the likelihood of achieving the full benefits of IT reform.

OMB has not established performance measures for evaluating the results of most of the IT reform initiatives GAO reviewed. Specifically, OMB has established performance measures for 4 of the 10 action items, including data center consolidation and cloud computing. However, no performance measures exist for 6 other action items, including establishing the best practices collaboration platform and developing a cadre of IT acquisition professionals. Until outcome-oriented performance measures are in place for each of the action items, OMB will be limited in its ability to evaluate progress that has been made and to determine whether or not the initiative is achieving its intended results.

GAO Assessment of Selected IT Reform Plan Action Items

Action Item	Status	Action Item	Status
Data center consolidation	◐	Guidance on modular development	◐
Cloud computing	◐	Budget models for modular development	◐
Contract vehicle for infrastructure	●	Routine IT purchases under agency CIO	◐
Best practices platform	◐	Investment review boards	●
IT acquisition professionals	●	Role of agency CIO and CIO Council	◐

Source: GAO analysis of OMB and agency data. Key: ● = Completed, ◐ = Partially completed

United States Government Accountability Office

Contents

Letter		1
	Background	2
	OMB and Key Federal Agencies Have Made Progress on IT Reform Action Items, But Much Remains to Be Done	14
	OMB and Key Agencies Plan to Address Items That We Found to Be Behind Schedule, But Lack Defined Time Frames for Completing Them	18
	OMB Has Not Established Measures for Evaluating Results on Most IT Reform Initiatives	20
	Conclusions	23
	Recommendations for Executive Action	23
	Agency Comments and Our Evaluation	24
Appendix I	Objectives, Scope, and Methodology	30
Appendix II	Comments from the Office of Management and Budget	33
Appendix III	Comments from the Department of Homeland Security	35
Appendix IV	Comments from the Department of Justice	37
Appendix V	Comments from the Department of Veterans Affairs	38
Appendix VI	GAO Contact and Staff Acknowledgments	40
Related GAO Products		41

Tables

	Table 1: OMB's IT Reform Plan: Action Items, Required Activities, and Responsible Parties	6

Table 2: GAO's Assessment of the Status of Key Action Items 15
Table 3: Assessment of Performance Measures Associated with
    Selected IT Reform Action Items 21

Abbreviations

CIO	Chief Information Officer
GSA	General Services Administration
IT	information technology
NIST	National Institute of Standards and Technology
OFPP	Office of Federal Procurement Policy
OMB	Office of Management and Budget
OPM	Office of Personnel Management

This is a work of the U.S. government and is not subject to copyright protection in the United States. The published product may be reproduced and distributed in its entirety without further permission from GAO. However, because this work may contain copyrighted images or other material, permission from the copyright holder may be necessary if you wish to reproduce this material separately.

United States Government Accountability Office
Washington, DC 20548

April 26, 2012

The Honorable Joseph I. Lieberman
Chairman
The Honorable Susan M. Collins
Ranking Member
Committee on Homeland Security and Governmental Affairs
United States Senate

The Honorable Thomas R. Carper
Chairman
Subcommittee on Federal Financial Management, Government
    Information, Federal Services, and International Security
Committee on Homeland Security and Governmental Affairs
United States Senate

In December 2010, the Federal Chief Information Officer (CIO) released a 25-point plan for reforming federal information technology (IT) management.[1] This document established an ambitious plan for achieving operational efficiencies and effectively managing large-scale IT programs. It also clearly identified actions to be completed in three different time frames: (1) within 6 months (by June 2011), (2) between 6 and 12 months (by December 2011), and (3) between 12 and 18 months (by June 2012).

To understand how agencies are implementing the IT Reform Plan, you asked us to (1) evaluate the progress the Office of Management and Budget (OMB) and key federal agencies have made on selected action items in the IT Reform Plan, (2) assess the plans for addressing any action items that are behind schedule, and (3) assess the extent to which sound measures are in place to evaluate the success of the IT reform initiatives.

To do so, we selected 10 action items from the IT Reform Plan, focusing on action items that (1) were expected to be completed by December 2011, (2) covered multiple different topic areas, and (3) were considered

---

[1]The Federal Chief Information Officer is a position within the Office of Management and Budget.

by internal and OMB subject matter experts to be the more important items. We also selected three federal agencies (the Departments of Homeland Security, Justice, and Veterans Affairs) based on several factors, including high levels of IT spending and large numbers of investments in fiscal year 2011. We then evaluated the steps OMB and the three federal agencies had taken to implement the selected action items from the IT Reform Plan. In cases where the action was behind schedule, we compared plans for addressing the schedule shortfalls to sound project planning practices.[2] We also determined whether and how agencies were tracking performance measures associated with these action items, and compared these measures to best practices in IT performance management.[3] In addition, we interviewed OMB and selected agency officials regarding progress, plans, and measures.

We conducted this performance audit from August 2011 to April 2012 in accordance with generally accepted government auditing standards. Those standards require that we plan and perform the audit to obtain sufficient, appropriate evidence to provide a reasonable basis for our findings and conclusions based on our audit objectives. We believe that the evidence obtained provides a reasonable basis for our findings and conclusions based on our audit objectives. See appendix I for a complete description of our objectives, scope, and methodology.

Background

IT can enrich people's lives and improve organizational performance. For example, during the last two decades the Internet has matured from being a means for academics and scientists to communicate with each other to a national resource where citizens can interact with their government in many ways, such as by receiving services, supplying and obtaining information, asking questions, and providing comments on proposed rules.

---

[2]See Carnegie Mellon University's Software Engineering Institute, *Capability Maturity Model® Integration for Acquisition, Version 1.3* (CMMI-ACQ, V1.3) and Project Management Institute Inc., *A Guide to the Project Management Body of Knowledge (PMBOK® Guide)–Fourth Edition*, (Newtown Square, PA: 2008).

[3]See OMB, *Guide to the Program Assessment Rating Tool* (Washington, D.C.: January 2008); Department of the Navy, Office of the Chief Information Officer, *Guide for Developing and Using Information Technology (IT) Performance Measurements* (Washington, D.C.: October 2001); and General Services Administration, Office of Governmentwide Policy, *Performance-Based Management: Eight Steps To Develop and Use Information Technology Performance Measures Effectively* (Washington, D.C.: 1996).

While investments in IT have the potential to improve lives and organizations, some federally funded IT projects can—and have—become risky, costly, unproductive mistakes. As we have described in numerous reports and testimonies, federal IT projects too frequently incur cost overruns and schedule slippages while contributing little to mission-related outcomes. Further, while IT should enable government to better serve the American people, the federal government has not achieved expected productivity improvements—despite spending more than $600 billion on IT over the past decade.

Roles and Responsibilities for Federal IT Management

Over the last two decades, Congress has enacted several laws to assist agencies and the federal government in managing IT investments. Key laws include the Paperwork Reduction Act of 1995,[4] the Clinger-Cohen Act of 1996,[5] and the E-Government Act of 2002.[6] Also, the GPRA (Government Performance and Results Act) Modernization Act of 2010 includes IT management as a priority goal for improving the federal government.[7]

- *Paperwork Reduction Act of 1995.* The act specifies OMB and agency responsibilities for managing information resources, including the management of IT. Among its provisions, this law establishes agency responsibility for maximizing the value and assessing and managing the risks of major information systems initiatives. It also requires that OMB develop and oversee policies, principles, standards, and guidelines for federal agency IT functions, including periodic evaluations of major information systems.

- *Clinger-Cohen Act of 1996.* The act places responsibility for managing investments with the heads of agencies and establishes CIOs to advise and assist agency heads in carrying out this responsibility. Additionally, this law requires OMB to establish processes to analyze, track, and evaluate the risks and results of major capital investments

---

[4] 44 U.S.C. § 3501 et seq.

[5] 40 U.S.C. § 11101 et seq.

[6] The E-Government Act of 2002, Pub. L. No. 107-347 (Dec. 17, 2002).

[7] Pub. L. No. 111-352, 124 Stat. 3866 (2011). The GPRA (Government Performance and Results Act) Modernization Act of 2010 amends the Government Performance and Results Act of 1993, Pub. L. No. 103-62, 107 Stat. 285 (1993).

in information systems made by federal agencies and report to Congress on the net program performance benefits achieved as a result of these investments.

- *E-Government Act of 2002.* The act establishes a federal e-government initiative, which encourages the use of web-based Internet applications to enhance the access to and delivery of government information and services to citizens, business partners, employees, and agencies at all levels of government. The act also requires OMB to report annually to Congress on the status of e-government initiatives. In these reports, OMB is to describe the administration's use of e-government principles to improve government performance and the delivery of information and services to the public.

- *GPRA (Government Performance and Results Act) Modernization Act of 2010.* The act establishes a new framework aimed at taking a more crosscutting and integrated approach to focusing on results and improving government performance. It requires OMB, in coordination with agencies, to develop long-term, outcome-oriented goals for a limited number of crosscutting policy areas at least every four years. The act specifies that these goals should include five areas: financial management, human capital management, IT management, procurement and acquisition management, and real property management.[8] On an annual basis, OMB is to provide information on how these long-term crosscutting goals will be achieved.

As set out in these laws, OMB is to play a key role in helping federal agencies manage their investments by working with them to better plan, justify, and determine how much they need to spend on projects and how to manage approved projects. Within OMB, the Office of E-government and Information Technology, headed by the Federal CIO, directs the policy and strategic planning of federal IT investments and is responsible for oversight of federal technology spending. In addition, the Office of Federal Procurement Policy (OFPP) is responsible for shaping the policies and practices federal agencies use to acquire the goods and services they need to carry out their missions.

---

[8] 31 U.S.C. § 1120(a)(1)(B).

Agency CIOs are also expected to have a key role in IT management. Federal law, specifically the Clinger-Cohen Act, has defined the role of the CIO as the focal point for IT management, requiring agency heads to designate CIOs to lead reforms that would help control system development risks; better manage technology spending; and achieve real, measurable improvements in agency performance.

In addition, the CIO Council—comprised of the CIOs and Deputy CIOs of 28 agencies and chaired by OMB's Deputy Director for Management—is the principal interagency forum for improving agency practices related to the design, acquisition, development, modernization, use, sharing, and performance of federal information resources. The CIO Council is responsible for developing recommendations for overall federal IT management policy, sharing best practices, including the development of performance measures, and identifying opportunities and sponsoring cooperation in using information resources.

Federal IT Reform Plan Strives to Address Persistent Challenges

After assessing the most persistent challenges in acquiring, managing, and operating IT systems, in December 2010, the Federal CIO established a 25-point IT Reform Plan designed to address challenges in IT acquisition, improve operational efficiencies, and deliver more IT value to the American taxpayer.[9] The actions were planned to be completed in three different time frames: (1) within 6 months (by June 2011), (2) between 6 and 12 months (by December 2011), and (3) between 12 and 18 months (by June 2012). Several different organizations were assigned ownership of the key action items, including the Federal CIO, the CIO Council, the General Services Administration (GSA), Office of Personnel Management (OPM), OFPP, the Small Business Administration, and other federal agencies. Table 1 contains detailed information on the action items in the IT Reform Plan. Shaded items are those selected for review in this report.

---

[9]OMB, *25 Point Implementation Plan to Reform Federal Information Technology Management,* (Washington, D.C.: Dec. 9, 2010).

Table 1: OMB's IT Reform Plan: Action Items, Required Activities, and Responsible Parties

Plan number	Action item title	Required activities	Responsible parties	Due date
1	Complete detailed implementation plans to consolidate 800 data centers by 2015	• Complete consolidation plans that include a technical roadmap, clear consolidation targets, and measurable milestones • Identify dedicated agency-specific program managers • Establish a cross-government task force comprised of the agency program managers • Ensure the task force meets monthly • Launch a public dashboard for tracking progress towards closures	OMB and federal agencies	June 2011
2	Create a governmentwide marketplace for data center availability	• Establish a governmentwide marketplace for agencies to market or obtain data center services	OMB and GSA	June 2012
3	Shift to a "cloud first" policy	• Establish a federal strategy for moving to cloud computing[a] • Identify three services (per agency) that are to move to cloud computing • Establish migration plans for the three services that are to move • Fully migrate the first service within 12 months	OMB and federal agencies	June 2011
4	Stand-up contract vehicles for secure Infrastructure-as-a-Service[b] solutions	• Make a common set of contract vehicles for secure cloud-based infrastructure solutions available governmentwide	GSA	June 2011
5	Stand-up contract vehicles for commodity services[c]	• Make contract vehicles for cloud-based e-mail solutions available governmentwide	GSA	December 2011
6	Develop a strategy for shared services	• Establish a vision and benchmarks for sharing services among federal agencies	CIO Council	December 2011
7	Design a formal IT program management career path	• Create a specialized career path for IT program managers that focuses on experience and expertise • Provide agencies authority to directly hire IT program managers • Have agencies identify and plan to fill competency gaps in IT program management	OPM and OMB	June 2011
8	Scale IT program management career path	• Expand IT program management career paths across the federal government	OPM and federal agencies	June 2012
9	Require integrated program teams	• Issue guidance requiring integrated program teams for all major IT programs • Dedicate resources throughout the program • Make program team members accountable for individual goals and overall program success	OMB	June 2011

Plan number	Action item title	Required activities	Responsible parties	Due date
10	Launch a best practices collaboration platform	• Establish a portal for program managers to exchange information on best practices • Require agencies to submit their experiences to the portal • Codify and synthesize agency submissions to provide a searchable database that facilitates real-time problem solving	CIO Council	June 2011
11	Launch technology fellows program	• Create a technology fellows program and the accompanying recruiting infrastructure to allow the government to tap into the emerging talent pool at universities with well-recognized technology programs	CIO Council	December 2011
12	Enable IT program manager mobility across government and industry	• Develop a process to encourage the movement of program managers across the government and industry in order to leverage their knowledge and expertise • Design opportunities for federal program managers to rotate through industry • Establish a repository of information on all federal government IT program managers	OMB, CIO Council, and OPM	June 2012
13	Design a cadre of specialized IT acquisition professionals[d]	• Define an IT acquisition specialist position • Establish the requirements, guidance, curriculum, and process for becoming one • Create guidance to strengthen the IT acquisition skills and capabilities of IT acquisition specialists	OMB and federal agencies	June 2011
14	Identify IT acquisition best practices and adopt governmentwide	• Study the experience of agencies with specialized IT acquisition teams • Develop a model for implementing such teams governmentwide	OFPP	June 2011
15	Issue contracting guidance and templates to support modular development	• Work with IT and acquisition community to develop guidance on contracting for modular development[e] • Obtain feedback from industry leaders • Develop templates and samples supporting modular development	OFPP	December 2011
16	Reduce barriers to entry for small innovative technology companies	• Take steps to develop clearer and more comprehensive policies for government contracting with small businesses	Small Business Administration, GSA, and OFPP	June 2012
17	Work with Congress to create IT budget models that align with modular development	• Analyze working capital funds and transfer authorities to identify current IT budget flexibilities • Identify programs at agencies where additional budget flexibilities could improve outcomes • Work with Congress to propose budgetary models to complement the modular development approach • Evaluate mechanisms for increased transparency for these programs	OMB and federal agencies	June 2011

Plan number	Action item title	Required activities	Responsible parties	Due date
18	Develop supporting materials and guidance for flexible IT budget models	• Develop a set of best practices and materials that explain the need for budget flexibilities and prescribe a path to achieve more flexible budget models	OMB, Chief Financial Officer Council, and CIO Council	December 2011
19	Work with Congress to scale flexible IT budget models more broadly	• Work with Congress to launch flexible IT budget models at selected agencies • Extend these budget models across the government as successes are demonstrated	OMB and federal agencies	June 2012
20	Work with Congress to consolidate commodity IT spending under agency CIO	• Work with Congress to consolidate commodity IT spending under the agency CIO • Develop a workable funding model for "commodity" IT services • Have the CIO Council and agency CIOs identify "commodity" services to be included in this funding model as they are migrated towards shared services	OMB and federal agencies	June 2011
21	Reform and strengthen Investment Review Boards	• Revamp IT budget submissions • Have agencies conduct "TechStat" reviews[f] • Have OMB analysts provide training to agency CIOs in the "TechStat" methodology	OMB and federal agencies	June 2011
22	Redefine role of agency CIOs and CIO Council	• Make agency CIOs responsible for managing the portfolio of large IT projects within their agencies • Have the CIO Council periodically review the highest priority "TechStat" findings assembled by the agency CIOs	Federal CIO and agency CIOs	June 2011
23	Roll out "TechStat" model at a component level	• Have an agency's component organizations conduct "TechStat" reviews • Make agency CIOs responsible for deploying the necessary tools and training on how to conduct "TechStat" reviews	Agency CIOs	June 2012
24	Launch "myth-busters" education campaign	• Identify the major myths that hinder the acquisition process • Reach out to key stakeholders in industry and government to dispel the myths	OFPP	June 2011
25	Launch an interactive platform for agency-industry collaboration before requests for proposals are issued	• Launch a governmentwide, online, interactive platform for identifying inexpensive, efficient solutions in the period prior to issuing a request for proposals	GSA	June 2011

Source: GAO analysis of OMB's IT Reform Plan.

Note: The shaded items are those reviewed in this report.

[a] Cloud computing is an emerging form of computing where users have access to scalable, on-demand capabilities that are provided through Internet-based technologies. It has the potential to provide IT services more quickly and at a lower cost.

[b] Infrastructure-as-a-Service is one type of cloud computing in which a vendor offers various infrastructure components such as hardware, storage, and other fundamental computing resources.

[c] Commodity services are systems or services used to carry out routine tasks (e.g., e-mail, data centers, and web infrastructure).

[d] While the IT Reform Plan discusses having agencies develop cadres of specialists, there is no requirement for agencies to do so.

eAccording to the IT Reform Plan, modular development is a system development technique that delivers functionality in shorter time frames by creating requirements at a high level and then refining them through an iterative process, with extensive engagement and feedback from stakeholders.

fOMB defines a TechStat as a face-to-face, evidence-based accountability review of an IT investment that results in concrete actions to address weaknesses and reduces wasteful spending by turning around troubled programs and terminating failed programs.

GAO Has Previously Reported on Needed Improvements in Federal IT Management

Given the challenges that federal agencies have experienced in acquiring and managing IT investments, we have issued a series of reports aimed at improving federal IT management over the last decade. Our reports cover a variety of topics, including data center consolidation, cloud computing, CIO responsibilities, system acquisition challenges, and modular development. Key reports that address topics covered in the IT Reform Plan include:

- *Data center consolidation.* In July 2011, we reported on the status of OMB's federal data center consolidation initiative.[10] Under this initiative, OMB required 24 participating agencies to submit data center inventories and consolidation plans by the end of August 2010. However, we found that only one of the agencies submitted a complete data center inventory and no agency submitted a complete data center consolidation plan. We concluded that until these inventories and plans are complete, agencies might not be able to implement their consolidation activities and realize expected cost savings. We recommended that agencies complete the missing elements in their plans and inventories. In response to our recommendations, in October and November 2011, the agencies updated their inventories and plans. We have ongoing work assessing the agencies' revised plans, and in February 2012, we reported that our preliminary assessment of the updated plans showed that not all agency plans were updated to include all required information.[11]

- *Cloud computing.* In May 2010, we reported on multiple agencies' efforts to ensure the security of governmentwide cloud computing. We noted that while OMB, GSA, and the National Institute of Standards and Technology (NIST) had initiated efforts to ensure secure cloud

---

[10]GAO, *Data Center Consolidation: Agencies Need to Complete Inventories and Plans to Achieve Expected Savings*, GAO-11-565 (Washington, D.C.: July 19, 2011).

[11]GAO, *Follow-up on 2011 Report: Status of Actions Taken to Reduce Duplication, Overlap, and Fragmentation, Save Tax Dollars, and Enhance Revenue*, GAO-12-453SP (Washington, D.C.: Feb. 28, 2012).

computing, significant work remained to be completed.[12] For example, OMB had not yet finished a cloud computing strategy; GSA had begun a procurement for expanding cloud computing services, but had not yet developed specific plans for establishing a shared information security assessment and authorization process; and NIST had not yet issued cloud-specific security guidance. We made several recommendations to address these issues. Specifically, we recommended that OMB establish milestones to complete a strategy for federal cloud computing and ensure it addressed information security challenges. OMB subsequently published a strategy which addressed the importance of information security when using cloud computing, but did not fully address several key challenges confronting agencies. We also recommended that GSA consider security in its procurement for cloud services, including consideration of a shared assessment and authorization process. GSA has since developed an assessment and authorization process for systems shared among federal agencies. Finally, we recommended that NIST issue guidance specific to cloud computing security. NIST has since issued multiple publications which address such guidance.

More recently, in October 2011, we testified that 22 of 24 major federal agencies reported that they were either concerned or very concerned about the potential information security risks associated with cloud computing.[13] These risks include being dependent on the security practices and assurances of vendors and the sharing of computing resources. We stated that these risks may vary based on the cloud deployment model. Private clouds, whereby the service is set up specifically for one organization, may have a lower threat exposure than public clouds, whereby the service is available to any paying customer. Evaluating this risk requires an examination of the specific security controls in place for the cloud's implementation.

We also reported that the CIO Council had established a cloud computing Executive Steering Committee to promote the use of cloud computing in the federal government, with technical and

---

[12]GAO, *Information Security: Federal Guidance Needed to Address Control Issues with Implementing Cloud Computing*, GAO-10-513 (Washington, D.C.: May 27, 2010).

[13]GAO, *Information Security: Additional Guidance Needed to Address Cloud Computing Concerns*, GAO-12-130T (Washington, D.C.: Oct. 5, 2011).

administrative support provided by GSA's Cloud Computing Program Management Office, but had not finalized key processes or guidance. A subgroup of this committee had developed the Federal Risk and Authorization Management Program, a governmentwide program to provide joint authorizations and continuous security monitoring services for all federal agencies, with an initial focus on cloud computing. The subgroup had worked with its members to define interagency security requirements for cloud systems and services and related information security controls.

- *Best practices in IT acquisition.* In October 2011, we reported on best practices in IT acquisitions in the federal government.[14] Specifically, we identified nine factors critical to the success of three or more of seven IT investments.[15] The factors most commonly identified include active engagement of stakeholders, program staff with the necessary knowledge and skills, and senior department and agency executive support for the program. We reported that while these factors will not necessarily ensure that federal agencies will successfully acquire IT systems because many different factors contribute to successful acquisitions, they may help federal agencies address the well-documented acquisition challenges they face.

- *IT spending authority.* In February 2008, we reported that the Department of Veterans Affairs had taken important steps toward a more disciplined approach to ensuring oversight of and accountability for the department's IT budget and resources.[16] These steps included providing the department's CIO responsibility for ensuring that there are controls over the budget and for overseeing all capital planning and execution, and designating leadership to assist in overseeing functions such as portfolio management.

---

[14] GAO, *Information Technology: Critical Factors Underlying Successful Major Acquisitions*, GAO-12-7 (Washington, D.C.: Oct. 21, 2011).

[15] The seven IT investments were identified by department officials as successful acquisitions in that they best achieved their respective cost, schedule, scope, and performance goals.

[16] GAO, *Information Technology: VA Has Taken Important Steps to Centralize Control of Its Resources, but Effectiveness Depends on Additional Planned Actions*, GAO-08-449T (Washington, D.C.: Feb. 13, 2008).

- *Investment review and oversight.* During the past several years, we issued numerous reports and testimonies on OMB's initiatives to highlight troubled IT projects.[17] We made multiple recommendations to OMB and federal agencies to enhance the oversight and transparency of federal IT projects. For example, in 2005 we recommended that OMB develop a central list of projects and their deficiencies, and analyze that list to develop governmentwide and agency assessments of the progress and risks of the investments, identifying opportunities for continued improvement.[18] In 2006, we recommended that OMB develop a single aggregate list of high-risk projects and their deficiencies and use that list to report to Congress on progress made in correcting high-risk problems.[19] As a result, OMB started publicly releasing aggregate data on its internal list of mission-critical projects that needed to improve (called its Management Watch List) and disclosing the projects' deficiencies. The agency also established a High-Risk List, which consisted of projects identified as requiring special attention from oversight authorities and the highest levels of agency management.

In June 2009, to further improve the transparency and oversight of agencies' IT investments, OMB publicly deployed a website, known as the IT Dashboard, which replaced its Management Watch List and High-Risk List. The data in the IT Dashboard is drawn from federal agencies' budget submissions.[20] OMB analysts use the IT Dashboard to identify IT investments that are experiencing performance problems and to select them for a TechStat session—a review of selected IT investments between OMB and agency leadership that is led by the

---

[17]GAO, *Information Technology: Management and Oversight of Projects Totaling Billions of Dollars Need Attention,* GAO-09-624T (Washington, D.C.: Apr. 28, 2009).

[18]GAO, *Information Technology: OMB Can Make More Effective Use of Its Investment Reviews,* GAO-05-276 (Washington, D.C.: Apr. 15, 2005).

[19]GAO, *Information Technology: Agencies and OMB Should Strengthen Processes for Identifying and Overseeing High Risk Projects,* GAO-06-647 (Washington, D.C.: June 15, 2006).

[20]Two different budget submissions, called exhibit 53s and exhibit 300s, provide the data accessible through the IT Dashboard. Exhibit 53s list all of the IT investments and their associated costs within a federal organization. An Exhibit 300, also called the Capital Asset Plan and Business Case, is used to justify resource requests for major IT investments and is intended to enable an agency to demonstrate, to its own management and to OMB, that a major investment is well planned.

Federal CIO. We have since completed three successive reviews of the data on the IT Dashboard and reported that while it is an important tool for reporting and monitoring major IT projects, the cost and schedule ratings were not always accurate for selected agencies.[21] We made recommendations to improve the accuracy of the data and, in our most recent report, found that the accuracy had improved.

In addition, in September 2011, we reported that OMB provides guidance to agencies on how to report on their IT investments, but this guidance does not ensure complete reporting or facilitate the identification of duplicative investments.[22] We recommended that OMB clarify its reporting on IT investments and improve its guidance to agencies on identifying and categorizing IT investments. OMB did not agree that further efforts were needed to clarify reporting and has not yet addressed our recommendations. Given the importance of continued improvement in OMB's reporting and guidance, we maintain that the recommendations are warranted.

- *Agency CIO responsibilities.* In September 2011, we reported that the responsibilities of the CIOs differ among agencies, and that CIOs face limitations in exercising their influence in certain IT management areas.[23] Specifically, CIOs do not always have sufficient control over IT investments, and they often have limited influence over the IT workforce, such as in hiring and firing decisions and the performance of component-level CIOs. We noted that more consistent implementation of CIOs' authority could enhance their effectiveness in these areas and that while OMB had taken steps to increase CIOs' effectiveness, it had not established measures of accountability to ensure that responsibilities are fully implemented. We recommended

---

[21] GAO, *IT Dashboard: Accuracy Has Improved, and Additional Efforts Are Under Way To Better Inform Decision Making*, GAO-12-210 (Washington, D.C.: Nov. 7, 2011); *Information Technology: OMB Has Made Improvements to Its Dashboard, but Further Work Is Needed by Agencies and OMB to Ensure Data Accuracy*, GAO-11-262 (Washington, D.C.: Mar. 15, 2011); and *Information Technology: OMB's Dashboard Has Increased Transparency and Oversight, but Improvements Needed*, GAO-10-701 (Washington, D.C.: July 16, 2010).

[22] GAO, *Information Technology: OMB Needs to Improve Its Guidance on IT Investments*, GAO-11-826 (Washington, D.C.: Sept. 29, 2011).

[23] GAO, *Federal Chief Information Officers: Opportunities Exist to Improve Role in Information Technology Management*, GAO-11-634 (Washington, D.C.: Sept. 15, 2011).

that OMB update its guidance to establish measures of accountability for ensuring that CIOs' responsibilities are fully implemented and require agencies to establish internal processes for documenting lessons learned. OMB officials generally agreed with our recommendations and, in August 2011, issued a memo to agencies emphasizing the CIO's role in driving the investment review process and responsibility over the entire IT portfolio for an agency.[24] The memo identified four areas in which the CIO should have a lead role: IT governance, program management, commodity services, and information security.

OMB and Key Federal Agencies Have Made Progress on IT Reform Action Items, But Much Remains to Be Done

OMB and key federal agencies have made progress on selected action items identified in the IT Reform Plan, but there are several areas where more remains to be done. Of the 10 key action items we reviewed, 3 were completed and the other 7 were partially completed by December 2011. The action items that are behind schedule share a common reason for the delays: the complexity of the initiatives. In all seven of the cases, OMB and the federal agencies are still working on the initiatives.

In a December 2011 progress report on its IT Reform Plan, OMB reported that it made greater progress than we determined. The agency reported that of the 10 action items, 7 were completed and 3 were partially completed. OMB officials from the Office of E-government and Information Technology explained that the reason for the difference in assessments is that they believe that the IT Reform Plan has served its purpose in acting as a catalyst for a set of broader initiatives. They noted that work will continue on all of the initiatives even after OMB declares the related action items to be completed under the IT Reform Plan. We disagree with this approach. In prematurely declaring the action items to be completed, OMB risks losing momentum on the progress it has made to date.

Table 2 provides both OMB's and our assessments of the status of the key action items, with action items rated as "completed" if all of the required activities identified in the reform plan were completed, and

---

[24]OMB, *Memorandum for Heads of Executive Departments and Agencies,* M-11-29 (Washington, D.C.: Aug. 8, 2011).

"partially completed" if some, but not all, of the required activities were completed.

Table 2: GAO's Assessment of the Status of Key Action Items

Plan number and action item title	OMB's reported status (as of December 2011)	GAO's assessment	Description
(1) Complete detailed implementation plans to consolidate at least 800 data centers by 2015	Completed	Partially completed	In 2011, agencies published their updated consolidation plans and identified dedicated program managers for their data center consolidation efforts. Also, OMB established a cross-government task force comprised of the agency program managers that meets monthly and launched a public dashboard for tracking progress in closing data centers. However, not all of the agencies' updated data center consolidation plans include the required elements. Of the three agencies we reviewed, one (the Department of Justice) lacked required milestones and targets for servers and utilization. In addition, in February 2012, we reported finding similar gaps in multiple agencies' consolidation plans.[a] When asked why the plans were not yet complete, agencies reported that it takes time to adequately plan for data center consolidation and many found that they need more time. We have previously recommended that agencies complete the missing elements from their data center consolidation plans.[b]
(3) Shift to cloud-first policy	Completed	Partially completed	The Federal CIO published a strategy for moving the government to cloud computing and had each agency identify three services to be moved to the cloud. In addition, each of the three agencies we reviewed established migration plans for these services and had migrated at least one service to the cloud by December 2011. However, each of the three agencies' migration plans we reviewed were missing key required elements, including a discussion of needed resources, migration schedules, or plans for retiring legacy systems. We have ongoing work performing a more detailed review of seven agencies' progress in implementing the federal cloud computing policy underway, and plan to issue that report in the summer of 2012.[c]
(4) Stand-up contract vehicles for secure Infrastructure-as-a-Service solutions	Completed	Completed	GSA has established a common set of contract vehicles for secure cloud-based infrastructure solutions, and made them available governmentwide. As of January 2012, federal agencies could purchase cloud solutions from three GSA-approved vendors.
(10) Launch a best practices collaboration platform	Completed	Partially completed	The CIO Council developed a web-based collaboration portal to allow program managers to exchange best practices and case studies, and all three agencies we reviewed have submitted case studies to OMB for the portal. However, the data accessible by the portal has not yet been effectively codified and synthesized, making it difficult for program managers to search the databases and for them to use it for problem solving. For example, a general search for cloud computing best practices identified more than 13,000 artifacts, while a date-bounded search for the last year identified 14 artifacts—of which only 8 clearly provided information on best practices in cloud computing. The vice chairman of the CIO Council explained that the portal's shortcomings are due to how new it is, and noted that the council is still working to improve the portal's functionality.

Plan number and action item title	OMB's reported status (as of December 2011)	GAO's assessment	Description
(13) Design a cadre of specialized IT acquisition professionals	Completed	Completed	In 2011, OFPP issued guidance defining an IT acquisition specialist; established the requirements, guidance, curriculum, and process for becoming one; and established guidance to strengthen the IT acquisition skills and capabilities of IT acquisition specialists. Because the development of the cadre is voluntary, the status of the agencies we reviewed varies: the Department of Veterans Affairs has a cadre of specialized IT acquisition professionals, the Department of Homeland Security is developing one, and the Department of Justice is still considering whether they need such a cadre.
(15) Issue contracting guidance and templates to support modular development	Partially completed	Partially completed	An OFPP official stated that the agency worked with the IT and acquisition community to develop draft guidance for modular development, and has obtained feedback from industry leaders. However, OFPP has not yet issued this guidance, or the required templates and samples supporting modular development. An OFPP official explained that delays were due to challenges in ensuring consistent definitions of modular development across the government and industry.
(17) Work with Congress to create IT budget models that align with modular development	Partially completed	Partially completed	OMB reported that it analyzed existing legal frameworks to determine what budget flexibilities are currently available and where additional budget flexibilities are needed, and worked to promote these ideas (such as multiyear budgets or revolving funds) with selected congressional committees. Also, the three agencies we reviewed identified programs where additional budget flexibilities could improve outcomes. For example, the Department of Homeland Security proposed a working capital fund for centralized IT operations and maintenance functions. However, in response to OMB's ideas, there has not yet been any new legislation to create budget authorities as a result of the IT Reform Plan and OMB has not identified options to increase transparency for programs that would fall under these budgetary flexibilities. OMB officials noted that they are behind schedule in working with Congress, in part because when the IT Reform Plan was issued in December 2010, the fiscal year 2012 budget process was already under way. They explained that this meant they needed to wait to incorporate changes into the fiscal year 2013 budget process.
(20) Work with Congress to consolidate commodity IT spending under agency CIOs	Partially completed	Partially completed	OMB issued a memo in August 2011 that, among other things, required agencies to consolidate commodity IT services under the agency CIO.[d] In addition, the federal CIO has discussed the importance of consolidating commodity IT under the agency CIOs with selected congressional committees. However, OMB noted that this action item is behind schedule and that it is continuing to discuss the implementation of the memo and the development of models for funding commodity IT with agencies and Congress. Further, the three agencies we reviewed had not yet reported to OMB on their proposals for migrating commodity IT services to shared services, in part because they were waiting for guidance from OMB. OMB officials noted that part of the reason for the delay is that when the IT Reform plan was issued in December 2010, the fiscal year 2012 budget process was already under way. Therefore, they needed to wait a year to incorporate changes into the fiscal year 2013 budget process.
(21) Reform and strengthen Investment Review Boards	Completed	Completed	In 2011, OMB revamped its requirements for agency IT budget submissions. OMB also developed, published, and provided training for agency CIOs on how to conduct TechStat reviews that includes accountability guidelines, engagement cadence, evaluation processes, and reporting processes. By December 2011, all 24 agencies conducted at least one TechStat review.

Plan number and action item title	OMB's reported status (as of December 2011)	GAO's assessment	Description
(22) Redefine role of agency CIOs and the CIO Council	Completed	Partially completed	In August 2011, OMB issued a memo directing agencies to strengthen the role of the CIO away from solely being responsible for policymaking and infrastructure maintenance to a role that encompasses true portfolio management for all IT. However, OMB acknowledged that there is disparity among agency CIOs' authorities and that it will take time for agencies to implement the required changes. Of the three agencies we reviewed, two CIOs reported having true portfolio management for all IT projects, and one did not. The Department of Homeland Security's CIO does not yet have responsibility for the portfolio of all IT projects. We have ongoing work assessing the Department's governance of IT investments. Regarding changes in the role of the CIO Council, the council formed a committee to focus on management best practices. This committee analyzed the outcomes of agency TechStat reviews over the past year and published a report discussing governmentwide trends in December 2011.

Source: GAO analysis of OMB and agency data.

[a]GAO-12-453SP.

[b]GAO-11-565.

[c]The seven agencies are the Departments of Agriculture, Health and Human Services, Homeland Security, State, and Treasury, as well as the General Services Administration and the Small Business Administration.

[d]OMB, Memorandum for Heads of Executive Departments and Agencies: Chief Information Officer Authorities, M-11-29 (Washington, D.C.: Aug. 8, 2011).

Until OMB and the agencies complete the action items called for in the IT Reform Plan, the benefits of the reform initiatives—including increased operational efficiencies and more effective management of large-scale IT programs—may be delayed. With the last of the action items in the IT Reform Plan due to be completed by June 2012, it will be important for OMB and the agencies to ensure that the action items due at earlier milestones are completed as soon as possible.

OMB and Key Agencies Plan to Address Items That We Found to Be Behind Schedule, But Lack Defined Time Frames for Completing Them

According to leading practices in industry and government, effective planning is critical to successfully managing a project. Effective project planning includes taking corrective actions when project deliverables fall behind schedule and defining time frames for completing the corrective actions.[25] As noted earlier in this report, we identified seven action items that are behind schedule or falling short of the IT Reform Plan's requirements. OMB and the agencies have plans for addressing all seven of the action items that we identified as behind schedule, but lack time frames for completing five of them. The seven action items we identified are:

- *Data center consolidation.* We noted that agencies' data center consolidation plans do not include all required elements. In July 2011, OMB directed agencies to complete the missing elements in their plans. The agencies are expected to provide an update on their plans in September 2012.

- *Cloud-first policy.* We noted that agencies' migration plans were missing selected elements. An OMB official stated while OMB did not review the quality of agency migration plans in order to close the reform plan action item, the official responsible for the cloud-first initiative would continue to work with agencies to ensure that the initiative was successful. There are no time frames for agencies to complete their migration plans.

- *Best practice collaboration portal.* We found that the best practices collaboration platform is missing key features that would allow the information to be accessible and useable. A CIO Council official noted that the council plans to improve the portal over time by adding the ability to load artifacts, allow users to chat online, contain an expertise repository, and allow or encourage labeling of information to improve the search for artifacts within the platform. However, the CIO Council has not established a time frame for providing additional functionality to the web-based collaboration portal.

- *Guidance and templates for modular contracting.* OFPP has not issued guidance or the required templates and samples supporting

---

[25]See Carnegie Mellon University's Software Engineering Institute, *Capability Maturity Model® Integration for Acquisition, Version 1.3 (CMMI-ACQ, V1.3)* and Project Management Institute Inc., *A Guide to the Project Management Body of Knowledge (PMBOK® Guide) – Fourth Edition*, (Newtown Square, PA: 2008).

modular development. It plans to continue developing guidance and templates to support modular development, and the first draft of this guidance is currently undergoing initial review. OFPP plans to issue its guidance and templates in spring 2012.

- *Obtaining new IT budget authorities.* OMB is behind schedule in obtaining new IT budget authorities. OMB officials stated that it plans to propose new authorities as part of the 2013 President's Budget, and intends to work with congressional committees throughout the budget rollout process. However, OMB has not yet established time frames for completing this activity.

- *Consolidating commodity IT under the agency CIO.* OMB is behind schedule in consolidating commodity IT spending under agency CIOs. OMB plans to propose new spending models for commodity IT in the 2013 President's Budget, and to work with Congress to implement these new models. However, OMB has not established a time frame for completing this activity.

- *Redefining roles of agency CIOs and the CIO Council.* OMB acknowledges that not all agency CIOs have authority for a full portfolio of IT investments and plans to collect data from agencies during spring 2012 to determine the extent to which the CIOs have this authority. At that point, OMB should be better positioned to determine what more needs to be done to ensure CIO roles are redefined. However, there is no time frame for completing this activity.

Until OMB and the agencies establish time frames for completing corrective actions, they increase the risk that key actions will not be effectively managed to closure. For example, without cloud migration plans, agencies risk maintaining legacy systems long after the system has been replaced by one operating in the cloud. Further, these incomplete actions reduce the likelihood of achieving the full range of benefits promised by the IT reform initiatives.

OMB Has Not Established Measures for Evaluating Results on Most IT Reform Initiatives

The importance of performance measures for gauging the progress of programs and projects is well recognized. In the past, OMB has directed agencies to define and select meaningful outcome-based performance measures that track the intended results of carrying out a program or activity.[26] Additionally, as we have previously reported, aligning performance measures with goals can help to measure progress toward those goals, emphasizing the quality of the services an agency provides or the resulting benefits to users.[27] Furthermore, industry experts describe performance measures as necessary for managing, planning, and monitoring the performance of a project against plans and stakeholders' needs.[28] According to government and industry best practices, performance measures should be measurable, outcome-oriented, and actively tracked and managed.

Recognizing the importance of performance measurement, OMB and GSA have established measures for 4 of the 10 action items we reviewed: data center consolidation, shifting to cloud computing, using contract vehicles to obtain Infrastructure-as-a-Service, and reforming investment review boards. Moreover, OMB reported on three of these measures in the analytical perspectives associated with the President's fiscal year 2013 budget. Specifically, regarding data center consolidation, OMB reported that agencies were on track to close 525 centers by the end of 2012 and expected to save $3 billion by 2015. On the topic of cloud computing, OMB reported that agencies had migrated 40 services to cloud computing environments in 2011 and expect to migrate an additional 39 services in 2012. Regarding investment review boards, OMB reported that agency CIOs held 294 TechStat reviews and had achieved more than $900 million in cost savings, life cycle cost avoidance, or reallocation of funding.

However, OMB has not established performance measures for 6 of the 10 action items we reviewed. For example, OMB has not established

---

[26] OMB, *Guide to the Program Assessment Rating Tool.*

[27] GAO, *NextGen Air Transportation System: FAA's Metrics Can Be Used to Report on Status of Individual Programs, but Not of Overall NextGen Implementation or Outcomes,* GAO-10-629 (Washington, D.C.: July 27, 2010).

[28] Thomas Wettstein and Peter Kueng, "*A Maturity Model for Performance Measurement Systems,*" and Karen J. Richter, Ph.D., Institute for Defense Analyses, *CMMI® for Acquisition (CMMI-ACQ) Primer, Version 1.2.*

measures related to the best practices collaboration platform, such as number of users, number of hits per query, and customer satisfaction. Further, while OMB has designed the guidance and curriculum for developing a cadre of IT acquisition professionals, it has not established measures for tracking agencies development of such a cadre. Table 3 details what performance measures and goals, if any, are associated with the action item.

Table 3: Assessment of Performance Measures Associated with Selected IT Reform Action Items

Action item	Performance measures	Performance goals
(1) Complete detailed implementation plans to consolidate 800 data centers by 2015	• Number of data center closures • Expected cost savings	• The IT Reform Plan identifies a goal to consolidate 800 data centers by 2015. • In December 2011, in conjunction with a decision to include smaller data centers in the consolidation effort, the Federal CIO increased this goal to more than 1000 data centers by 2015. • In February 2012, OMB announced a goal of saving $3 billion by 2015.
(3) Shift to a cloud-first policy	• Number of services transitioned to a cloud computing environment • Number of legacy systems eliminated • Anticipated cost savings	• The IT Reform Plan states that each agency will identify three services to move to the cloud and that one of those services must move within 12 months. • OMB has not yet announced goals for eliminated legacy systems or anticipated cost savings.
(4) Stand-up contract vehicles for secure Infrastructure-as-a-Service solutions	• Number of task orders issued under the contract vehicle • Dollar amounts awarded through the contract vehicle • Period of performance for the contract	• GSA established a goal of having at least one task order issued under the Infrastructure-as-a-Service blanket purchase agreement in the first year. • GSA has not yet announced goals for its second year.
(10) Launch a best practices collaboration platform	–[a]	–[a]
(13) Design a cadre of specialized IT acquisition professionals	–[a]	–[a]
(15) Issue contracting guidance and templates to support modular development	–[a]	–[a]
(17) Work with Congress to create IT budget models that align with modular development	–[a]	–[a]
(20) Work with Congress to consolidate commodity IT spending under agency CIOs	–[a]	–[a]

Action item	Performance measures	Performance goals
(21) Reform and strengthen investment review boards	• Number of TechStat reviews • Number of terminated programs • Cost savings associated with TechStat reviews	• OMB established a goal of having agency CIOs terminate or turn around one third of all underperforming IT investments by June 2012.
(22) Redefine role of agency CIOs and the CIO Council	—[a]	—[a]

Source: GAO analysis of OMB and agency data.

[a]Performance measures or goals have not been established for this action item.

OMB officials, including two policy analysts within the Office of E-government and Information Technology who are responsible for the IT Reform Plan, stated that they do not believe that it is appropriate for OMB to establish measures for the action items in the IT Reform Plan. The officials explained that they believe that the purpose of the IT Reform Plan is to act as a catalyst for initiatives that are expected to continue outside of the plan. For example, the IT Reform Plan called for OMB and agencies to complete several discrete activities to push forward on data center consolidation, but the Federal Data Center Consolidation Initiative will continue on well after the deliverables noted in the reform plan are completed. They acknowledged that it would be appropriate to have performance measures for each of the broader initiatives outside of the IT Reform Plan, but noted that this should be the responsibility of the group running each initiative.

We disagree with OMB's view and believe that performance measures are a powerful way to motivate people, communicate priorities, and improve performance. In our assessment, we sought any available performance measures associated with either the action item or the broader initiative, and in cases like the data center consolidation initiative, gave credit for the measures that were established for the initiative. However, we found that most action items and initiatives lacked any performance measures. Moreover, if OMB encourages individual agencies to establish measures, there will likely be multiple different measures for the action items and it would be more difficult to demonstrate governmentwide progress. Therefore, we believe that it is appropriate for OMB to establish performance measures for each of the action items in order to effectively measure the results of the IT Reform Plan. Until OMB establishes and begins tracking measurable, outcome-oriented performance measures for each of the action items, the agency will be limited it its ability to evaluate progress that has been made and whether or not the initiative is achieving its goals.

Conclusions

OMB and selected agencies have made strides in implementing the IT Reform Plan, including pushing agencies to consolidate data centers, migrating federal services to cloud computing, improving the skills of IT acquisition professionals, and strengthening the roles and accountability of CIOs. However, several key reform items are behind schedule and OMB lacks time frames for completing most of them. Despite reporting that selected actions are completed, OMB and federal agencies are still working on them. This sends an inconsistent message on the need to maintain focus on these important initiatives. Moving forward, it will be important for OMB to accurately characterize the status of the action items in the IT Reform Plan in order to keep agencies' focus and momentum on these important reform initiatives.

OMB has not established performance measures for gauging the success of most of its reform initiatives. For example, while OMB is tracking the number of services that agencies move to a cloud computing environment and the number of data center closures, it is not tracking the usefulness of its efforts to develop a best practices collaboration portal or a cadre of IT acquisition professionals.

Until OMB and the agencies complete the action items called for in the IT Reform Plan, establish time frames for completing corrective actions, and establish performance measures to track the results of the reform initiatives, the government may not be able to realize the full promise of the IT Reform Plan. The IT Reform Plan's goals of improving government IT acquisitions and the efficiency of government operations are both ambitious and important, and they warrant a more structured approach to ensure actions are completed and results are achieved.

Recommendations for Executive Action

To help ensure the success of IT reform initiatives, we are making four recommendations to OMB. Specifically we are recommending that the Director of the Office of Management and Budget direct the Federal Chief Information Officer to

- ensure that the action items called for in the IT Reform Plan are completed by the responsible parties prior to the completion of the IT Reform Plan's 18 month deadline of June 2012, or if the June 2012 deadline cannot be met, by another clearly defined deadline;

- provide clear time frames for addressing the shortfalls associated with the IT Reform Plan action items;

- accurately characterize the status of the IT Reform Plan action items in the upcoming progress report in order to keep momentum going on action items that are not yet completed; and

- establish outcome-oriented measures for each applicable action item in the IT Reform Plan.

We are also making two recommendations to the Secretaries of Homeland Security and Veterans Affairs and to the Attorney General of the Department of Justice to address action items in the IT Reform Plan where the agencies have fallen behind. Specifically, we are recommending that they direct their respective agency CIOs to

- complete elements missing from the agencies' plans for migrating services to a cloud computing environment, as applicable, and

- identify and report on the commodity services proposed for migration to shared services.

Agency Comments and Our Evaluation

We received comments on a draft of our report from OMB; the Departments of Homeland Security, Justice, and Veterans Affairs; and GSA. OMB agreed with two recommendations and disagreed with two recommendations; the Departments of Homeland Security, Justice, and Veterans Affairs generally agreed with our recommendations; and GSA did not agree or disagree with our recommendations. Each agency's comments are discussed in more detail below.

- OMB's Federal CIO provided written comments on a draft of this report, as well as supplementary comments via e-mail. The written comments are provided in appendix II. The Federal CIO stated that OMB believes our analysis and findings have been critical to driving IT reforms across the federal government, and that OMB plans to use this report to continue the positive momentum on the IT Reform Plan. In addition, the Federal CIO stated that despite agreeing with many of the observations and recommendations in the draft report, OMB had concerns with selected recommendations, observations, and the scope of our review. The agency's comments and, where applicable, our evaluation follow:

  - OMB agreed with our recommendation to ensure that action items called for in the IT Reform Plan are completed by the end of the IT Reform Plan's 18-month deadline of June 2012 and stated that

OMB intends to complete the action items by the deadline.

- OMB agreed with our recommendation to provide clear time frames for addressing the shortfalls associated with the IT Reform Plan action items and stated that OMB will provide clear time frames where applicable.

- OMB disagreed with our recommendation that the agency accurately characterize the status of IT Reform Plan action items in the upcoming progress report. The agency stated that it has accurately characterized the completeness of the action items, and therefore, the recommendation does not apply. As stated in this report, we do not agree with OMB's characterization of four action items: data center consolidation, cloud-first policy, best practices collaboration portal, and redefining roles of agency CIOs and the CIO Council. OMB considers these action items to be completed. We do not.

  While OMB has made progress in each of these areas, we found activities specified in the IT Reform Plan that have not yet been completed. Specifically, in the area of data center consolidation, we found that selected agency plans are still incomplete; in the move to cloud computing, selected agency migration plans lack key elements; in the area of the best practices portal, we found that the portal lacks key features that would allow the information to be accessible and useful to program managers; and in revising CIO roles, we identified an agency that does not yet have the envisioned authority over IT acquisitions. Further, in a recent memorandum to agency CIOs, the Federal CIO acknowledged that agency data center consolidation plans are incomplete and required agencies to provide an annual update to the plans.[29] In addition, our assessment that the cloud migration plans are incomplete was affirmed by the three agencies we reviewed agreeing with our recommendation that they complete cloud migration plans. Thus, we believe that our recommendation to OMB to accurately characterize the status of IT Reform action items is valid.

---

[29]OMB, *Memorandum for Chief Information Officers*, (Washington, D.C.: Mar. 19, 2012).

- OMB disagreed with our recommendation to establish outcome-oriented measures for each applicable action item in the IT Reform Plan, noting that the agency measured the completeness of the IT Reform actions and not the performance measures associated with broader initiatives. OMB also suggested that we erroneously gave the agency credit for performance measures associated with broader initiatives on data center consolidation, cloud computing, and investment review boards. We acknowledge that some of the action items in the IT Reform Plan are subsets of broader initiatives, and where applicable, we gave credit for having measures associated with the broader initiatives. We continue to believe that this approach is appropriate because the action items and the broader initiatives are intrinsically intertwined. For instance, it would have been unfair to state that there are no measures associated with consolidating federal data centers when such measures clearly exist.

  Moreover, the point remains that there are multiple action items in the IT Reform Plan that are not aligned with broader initiatives and for which there are no measures. Examples include the best practices portal, development of a cadre of specialized IT acquisition professionals, and establishing budget models that align with modular development. Given that the purpose of the IT Reform Plan is to achieve operational efficiencies and improve the management of large-scale IT programs, we continue to assert that it is appropriate to establish performance measures to monitor the IT Reform Plan's results. According to the administration's public website intended to provide a window on efforts to deliver a more effective, smarter, and leaner government, performance measurement is a necessary step in improving performance and that it helps set priorities, tailor actions, inform on progress, and diagnose problems.[30] Until OMB establishes and tracks measureable, outcome-oriented performance measures for each of the action items in the IT Reform Plan, the agency will be limited in its ability to evaluate progress that has been made and whether or not the initiative is achieving its goals.

- OMB stated that the title of our draft report (*Information Technology Reform: Progress is Mixed; More Needs to Be Done*

---

[30]See www.performance.gov.

to Complete Actions and Measure Results) did not accurately capture the substantial and overwhelmingly positive progress made to date. Moreover, OMB stated that the responsible entities have completed 81.5 percent of the required activities associated with the 10 action items we reviewed. We acknowledge the progress OMB and agencies have made on IT Reform Plan items in this report and have modified the title of our report to reflect that progress. However, our analysis of the percentage of completed activities differs from OMB's calculations. The 10 action items we reviewed include 31 distinct required activities (see table 1). We found that the responsible entities completed 18 of these activities—a 58 percent completion rate.

- OMB also stated that our assessment should acknowledge that OMB does not have the statutory authority to carry out certain action items without congressional action. These action items involved creating IT budget models to align with modular development and consolidating commodity IT spending under the agency CIOs. The Federal CIO stated that although OMB has taken steps to engage with Congress, the agency cannot unilaterally grant budget flexibilities or consolidate spending. While it is true that completing these items depends upon congressional action, according to the IT Reform Plan, it is the responsibility of OMB and the federal agencies to work with Congress to propose budget models to address these items.

- In general, OMB stated that it will continue to drive reform throughout the federal government via the completion of the remaining actions in the IT Reform Plan, as well as continuing to work with agencies as they implement broader initiatives such as data center consolidation and the transition to cloud computing.

- In supplementary comments provided via e-mail, the Federal CIO also expressed concerns with the scope of our report, stating that the intent of the IT Reform Plan was not to reform all federal IT, but to establish some early wins to garner momentum for OMB's broader initiatives. The Federal CIO also noted that OMB has been consistent in publicizing the IT Reform Plan as an 18-month plan with discrete goals designed to augment and accelerate broader initiatives that existed before the IT Reform Plan was launched and would continue after the plan has been completed.

We believe that the scope of our review is appropriate. Since its inception, the scope of our review has focused on the action items and supporting activities noted in the IT Reform Plan. All of the required activities listed in table 1 in the background section of this report are listed in the IT Reform Plan. Moreover, we did not evaluate activities that are outside of the IT Reform Plan, such as OMB's efforts to establish a cost model for agencies to use in estimating the costs and savings of data center consolidation. Further, we agree that to completely reform IT, OMB and agencies must undertake activities beyond the IT Reform Plan's 18-month time frame. The activities within the IT Reform Plan are essential building blocks that will carry on well beyond the IT Reform Plan's end.

- In written comments, the Department of Homeland Security's Director of Departmental GAO-Office of Inspector General Liaison Office concurred with our recommendations and identified steps that the agency is undertaking to address them. The department's written comments are provided in appendix III.

- In written comments, the Department of Justice's Assistant Attorney General for Administration generally agreed with our recommendations and identified steps that the agency has undertaken to address them. The department's written comments are provided in appendix IV.

- In written comments, the Chief of Staff at the Department of Veterans Affairs agreed with our recommendations and identified steps that the department is taking to implement them. The department's written comments are provided in appendix V.

- In comments provided via e-mail, a Management and Program Analyst within GSA's Office of Administrative Services stated that the agency had no official response or technical comments on the draft report.

As agreed with your offices, unless you publicly announce the contents of this report earlier, we plan no further distribution until 30 days from the report date. At that time, we will send copies to interested congressional committees, the secretaries and administrators of the departments and agencies addressed in this report, and other interested parties. In addition, the report will be available at no charge on the GAO website at http://www.gao.gov.

If you or your staffs have any questions on the matters discussed in this report, please contact me at (202) 512-9286 or pownerd@gao.gov. Contact points for our Offices of Congressional Relations and Public Affairs may be found on the last page of this report. GAO staff who made major contributions to this report are listed in appendix VI.

David A. Powner
Director, Information Technology
 Management Issues

Appendix I: Objectives, Scope, and Methodology

Our objectives were to (1) evaluate the progress the Office of Management and Budget (OMB) and key federal agencies have made on selected action items in the Information Technology (IT) Reform Plan, (2) assess the plans for addressing any action items that are behind schedule, and (3) assess the extent to which sound measures are in place to evaluate the success of the IT reform initiatives.

In establishing the scope of our engagement, we selected ten action items for review, focusing on action items that (1) were due at the 6 or 12 month milestones because these were expected to be completed during our review, (2) covered multiple different topic areas, and (3) were considered by internal and OMB subject matter experts to be the more important items. These action items are:

- Complete detailed implementation plans to consolidate 800 data centers by 2015.
- Shift to a "cloud first" policy.
- Stand-up contract vehicles for secure Infrastructure-as-a-Service solutions.
- Launch a best practices collaboration platform.
- Design a cadre of specialized IT acquisition professionals.
- Issue contracting guidance and templates to support modular development.
- Work with Congress to create IT budget models that align with modular development.
- Work with Congress to consolidate Commodity IT spending under agency Chief Information Officers (CIO).
- Reform and strengthen Investment Review Boards.
- Redefine the role of agency CIOs and the CIO Council.

In addition, in the seven cases where multiple agencies are identified as a responsible entity for the action item, we selected three civilian agencies (the Departments of Homeland Security, Veterans Affairs, and Justice) based on factors including (1) high levels of IT spending in fiscal year

Appendix I: Objectives, Scope, and Methodology

2011, (2) poor performance on the IT Dashboard, (3) high number of major IT investments in fiscal year 2011, and (4) coverage of agencies that were not included on other GAO reviews of IT reform initiatives.

To evaluate OMB and federal agencies progress in implementing the IT Reform Plan, we evaluated efforts by the entities responsible for each of the action items, including OMB, the General Services Administration (GSA), the Chief Information Officers (CIO) Council, and selected agencies. For each of the 10 action items in the IT Reform Plan, we reviewed OMB's guidance and identified required activities. We compared agency documentation to these requirements, and identified gaps and missing elements. We rated each action item as "completed" if the responsible agencies demonstrated that they completed the required activities on or near the due date, and "partially completed" if the agencies demonstrated that they completed part of the required activities. We interviewed agency officials to clarify our initial findings and to determine why elements were incomplete or missing.

To assess the plans for addressing any action items that are behind schedule, we identified the agencies' plans for addressing the schedule shortfalls and compared these to sound project planning practices identified by organizations recognized for their experience in project management and acquisition processes.[1] We also interviewed relevant agency officials regarding the reasons that their activities were behind schedule and the impact of any shortfalls in their mitigation plans.

To assess the extent to which sound measures are in place to evaluate success, we determined whether performance measures were applicable for each of the selected action items, and if so, how agencies were tracking these measures. We compared these measures to best practices in IT performance management identified by leading industry and government organizations[2] and assessed other options for measuring

---

[1] See Carnegie Mellon University's Software Engineering Institute, *Capability Maturity Model® Integration for Acquisition, Version 1.3 (CMMI-ACQ, V1.3)* and Project Management Institute Inc., *A Guide to the Project Management Body of Knowledge (PMBOK® Guide)–Fourth Edition*, (Newtown Square, PA: 2008).

[2] See OMB, *Guide to the Program Assessment Rating Tool*; Department of the Navy, *Guide for Developing and Using Information Technology (IT) Performance Measurements*; and General Services Administration, Office of Governmentwide Policy, *Performance-Based Management: Eight Steps To Develop and Use Information Technology Performance Measures Effectively*.

Appendix I: Objectives, Scope, and Methodology

performance. In addition, we interviewed OMB and selected agency officials regarding progress, plans, and measures. As we were completing our audit work, OMB reported making progress in its efforts to consolidate data centers, transition to a cloud computing environment, and strengthen investment review boards, and provided data on specific measures within each of these areas. We assessed the reliability of the data provided on these measures by obtaining information from agency officials and from the CIO Council regarding their efforts to ensure the reliability of the data. While we identified limitations in the quality of the data that agencies reported, we determined that this data was sufficiently reliable for the purpose of presenting a general overview of progress in establishing performance measures.

We conducted our work at multiple agencies' headquarters in the Washington, D.C., metropolitan area. We conducted this performance audit from August 2011 to April 2012 in accordance with generally accepted government auditing standards. Those standards require that we plan and perform the audit to obtain sufficient, appropriate evidence to provide a reasonable basis for our findings and conclusions based on our audit objectives. We believe that the evidence obtained provides a reasonable basis for our findings and conclusions based on our audit objectives.

Appendix II: Comments from the Office of Management and Budget

EXECUTIVE OFFICE OF THE PRESIDENT
OFFICE OF MANAGEMENT AND BUDGET
WASHINGTON, D.C. 20503

April 9, 2012

David Powner
Director, Information Technology Management Issues
The General Accountability Office
441 G St. NW
Washington D.C. 20548

Dear Mr. Powner:

Thank you for the opportunity to comment on the Government Accountability Office's (GAO) draft report, "Information Technology Reform: Progress Is Mixed; More Needs to Be Done to Complete Actions and Measure Results" (GAO-12-461). OMB believes that GAO's analyses and findings have been critical to driving IT reforms across the Federal Government. As with other reports, OMB plans to use this report to continue the positive momentum on the Information Technology (IT) Reform Plan.

OMB is in agreement with many of the observations and recommendations in the draft report, and we have already initiated actions to address the following items:

1) In general, OMB will continue to drive reform throughout the Federal Government, via completion of the remaining actions in the IT Reform Plan, as well as through continuing to work with Agencies as they implement initiatives such as Data Center Consolidation and Cloud First;

2) OMB acknowledges the delay in delivering *IT Reform Plan Action #15 – Issue contracting guidance and templates to support modular development*. This guidance is currently under development, and OMB plans issuance prior to June 2012.

3) OMB agrees with GAO's first recommendation to *"ensure that the action items called for in the IT Reform Plan are completed by the end of the IT Reform Plan's 18-month deadline of June 2012."* OMB intends to complete prescribed action items from the IT Reform Plan by June 2012.

4) OMB agrees with GAO's second recommendation to *provide clear timeframes for addressing the shortfalls associated with the IT Reform Plan Action Items*. OMB has revised plans to complete the prescribed action items by June 2012 and will provide clear timelines for completion where applicable.

Despite broad agreement on the findings in this draft, OMB requests clarification or re-examination of the following findings and recommendations:

1) The title of the draft report, "Information Technology Reform: Progress Is Mixed; More Needs to Be Done to Complete Actions and Measure Results" does not accurately capture the substantial and overwhelmingly positive progress made to-date. By GAO's measure, Agencies and OMB have completed 22 out of 27 (81.5%) required activities that were seemingly used to measure completeness – this does not denote "mixed progress" in our assessment. Additionally, the current title of the draft report broadens the scope of the engagement to encompass the broader topic of "Information Technology Reform", although this scope is not reflected consistently in the draft report;

2) OMB considers Action items #1, #3, #10 and #22 to be completed. OMB's position is that the IT Reform Plan outlined a set of discrete action items designed to augment and accelerated broader initiatives that existed before the IT Reform Plan was launched and will continue after the plan has finished. We consider the action items to be complete because the required activities associated with

Appendix II: Comments from the Office of Management and Budget

that action item have been completed. However, we will continue to measure the performance of agencies within the broader initiatives that have benefitted from IT Reform action items (such as Cloud First, Data Center, and CIO Authorities).

3) OMB disagrees with the need for the third recommendation to *"accurately characterize the status of the IT Reform Plan action items in the upcoming progress report in order to keep momentum going on actions items that are not yet completed."* OMB considers all IT Reform action items have been accurately characterized based on the stated intent of the IT Reform Plan, and therefore this recommendation does not apply. We have provided GAO justification for this position and request the completeness of each action item is appropriately categorized.

4) OMB also disagrees with the need for the fourth recommendation, *"...to establish outcome-oriented measures for each applicable action items in the IT Reform Plan"*, given the stated intent of the IT Reform Plan. We measured the completeness of IT Reform actions and not the performance measures associated with broader initiatives. Yet, GAO lists performance measures that have been pulled from elsewhere and are not a part of the IT Reform Plan itself;

5) OMB believes GAO's assessment should take into consideration the fact that OMB does not have the statutory authority to carry out certain action items without Congressional action. For example, although OMB has taken steps to engage with Congress, OMB cannot unilaterally grant budget flexibilities sought in action item #17. OMB requests that GAO makes such dependencies clear in its report.

In sum, we ask that GAO be consistent across IT Reform action items and their associated required activities. This would enable us to hold ourselves and Agencies accountable for truly incomplete actions within the IT Reform Plan.

The 25 Point Plan to Reform Federal Information Technology Management has been a unique and sometimes polarizing effort, but the effort has undoubtedly changed the landscape of Federal IT for the better. Agencies and OMB switched the default on numerous management efforts so that continuous improvement could carry on well beyond the 18 month timeframe that bookend the IT Reform Plan. OMB makes great use of well thought out and executed GAO reports in implementing priority initiatives and look forward to a more focused final report. Thank you for the opportunity to review and comment on this draft report.

Steven VanRoekel

Appendix III: Comments from the Department of Homeland Security

U.S. Department of Homeland Security
Washington, DC 20528

April 3, 2012

David A. Powner
Director, Information Technology Management Issues
441 G Street, NW
U.S. Government Accountability Office
Washington, DC 20548

Dear Mr. Powner:

Re: Draft Report GAO-12-461, "INFORMATION TECHNOLOGY REFORM: Progress Is Mixed; More Needs to Be Done to Complete Actions and Measure Results"

Thank you for the opportunity to review and comment on this draft report. The U.S. Department of Homeland Security (DHS) appreciates the U.S. Government Accountability Office's (GAO's) work in planning and conducting its review and issuing this report.

The Department is pleased to note GAO's positive acknowledgement of the actions agencies have taken in implementing the Information Technology (IT) Reform Plan, including consolidation of data centers and migrating federal services to cloud computing platforms. In particular, we appreciate GAO's recognition that action items behind schedule have this status because "these initiatives are complex."

The draft report contained two recommendations directed to the Secretaries of the Departments of Homeland Security, Justice, and Veteran's Affairs to address action items in the IT Reform Plan where the agencies have fallen behind. DHS concurs with both recommendations. Specifically, GAO recommended that the Secretaries direct their respective agency Chief Information Officers (CIOs) to:

Recommendation 1: Complete elements missing from the agencies' plan for migrating services to a cloud computing environment.

Response: Concur. In response to the Office of Management and Budget (OMB) "cloud first" policy, DHS Office of the Chief Information Officer (OCIO) personnel have drafted a strategic plan for moving to the cloud with commodity enterprise services, identified the services that are to move to the cloud, and established migration plans for the services. OCIO personnel will continue to mature the migration plans for these services and address the elements missing from the plans.

Recommendation 2: Identify and report on the commodity services proposed for migration to shared services.

Appendix III: Comments from the Department of Homeland Security

Response: Concur. DHS OCIO personnel intend to identify and provide updates to OMB on the commodity services proposed for migration. In March 2012, they reported two shared services to OMB (1) UniTrac, a unified scalable system that allows seamless integration of data from multiple sources using satellite and terrestrial communications systems, and (2) Workplace as a Service (WPaaS), an offering that will provide robust virtual desktop, remote access, and other mobile services.

In addition, OCIO personnel produce a DHS Cloud Scorecard which lists the private and public cloud commodities that are currently or soon-to-be available. To ensure full transparency, the scorecard provides service planning and design information, performance measures (dates for Authority to Operate and operational), status comments, and migration status information. The scorecard is updated as changes occur to the information being reported and distributed monthly to the CIO and quarterly to the DHS Under Secretary for Management.

Again, thank you for the opportunity to review and comment on this draft report. We look forward to working with you on future Homeland Security issues.

Sincerely,

Jim H. Crumpacker
Director
Departmental GAO-OIG Liaison Office

Appendix IV: Comments from the Department of Justice

U.S. Department of Justice

Washington, D.C. 20530

MAR 3 0 2012

Mr. David A. Powner
Director, Information Technology Management Issues
U.S. Government Accountability Office
441 G Street, NW
Washington, DC 20548

Dear Mr. Powner:

The Department of Justice has reviewed the Government Accountability Office's (GAO) draft report "Information Technology Reform: Progress is Mixed; More Needs to Be Done to Complete Actions and Measure Results (GAO-12-461)."

The Department appreciates the opportunity to review and comment on the GAO's draft report. We concur with the recommendations included in the report and will respond in detail to both recommendations in our 60-day response to Congress. In the interim, we would like to provide the following comments in response to GAO's recommendations:

Regarding the first recommendation to "complete elements missing from the agencies' plans for migrating services to a cloud computing environment," the Department receives periodic reports from the organizations responsible for the migration of the three "Must Move" cloud services. These reports include the respective project migration schedules, which the Department delivers to OMB as requested.

Regarding the second recommendation to "identify and report on the commodity services proposed for migration to shared services," subsequent to our interview with the GAO team, the Department identified the commodity services that will be migrated to Shared Services, and reported the information to OMB on March 1, 2012 in compliance with OMB's schedule.

Should you have any questions regarding this topic, please do not hesitate to contact Louise Duhamel, Acting Assistant Director, Audit Liaison Group on 202-514-0469.

Sincerely,

Lee J. Lofthus
Assistant Attorney General
for Administration

Appendix V: Comments from the Department of Veterans Affairs

DEPARTMENT OF VETERANS AFFAIRS
Washington DC 20420

March 29, 2012

Mr. David A Powner
Director, Information Technology
  Management Issues
U.S. Government Accountability Office
441 G Street, NW
Washington, DC 20548

Dear Mr. Powner:

The Department of Veterans Affairs (VA) has reviewed the Government Accountability Office's (GAO) draft report, *"Information Technology Reform: Progress is Mixed; More Needs to Be Done to Complete Actions and Measure Results"* (GAO-12-461) and is providing comments in the enclosure.

VA appreciates the opportunity to comment on your draft report.

Sincerely,

John R. Gingrich
Chief of Staff

Enclosure

Appendix V: Comments from the Department of Veterans Affairs

Enclosure

Department of Veterans Affairs (VA) Comments to
Government Accountability Office (GAO) Draft Report:
*"Information Technology Reform: Progress is Mixed; More
Needs to Be Done to Complete Actions and Measure Results"*
(GAO-12-461)

GAO Recommendation: To address action items in the IT Reform Plan where agencies have fallen behind, GAO recommends that agencies direct their CIO to:

Recommendation 1: complete elements missing from the agencies' plans for migrating services to a cloud computing environment.

VA Comment: Concur. VA will update its plan to include a discussion of needed resources, migration schedule, and plans for retiring legacy systems by June 30, 2012.

Recommendation 2: identify and report on the commodity services proposed for migration to shared services.

VA Comment: Concur. There are numerous commodity services cited in the Office of Management and Budget's Memo M-11-29 that VA has already implemented in a cost effective, enterprise-wide manner. Some examples include:

- Data Centers – All data center activities are managed by the National Data Center Program within the Enterprise Operations Division. VA is leveraging shared services in its National Data Center Program by co-locating systems in Defense Information Systems Agency Data Centers.
- Desktop Computers – A single enterprise contract for desktops was initially awarded 4 years ago and the replacement contract was awarded last year. All desktops for the enterprise are purchased via this vehicle and are all standard configurations with a single standard operating system image on them.
- Servers – Lifecycle management processes ensure standard server configurations are procured in enterprise-wide procurements. This drives efficiencies by reducing overall costs, simplifying the management of the systems, and reducing the administrative burden of the procurement process.
- Mobile Devices – All mobile devices are managed by the enterprise mobile device management tools and are standardized on Blackberry along with a pilot of iPhone/iPad devices.
- E-mail – VA has run a standard enterprise e-mail system since 1997 and continues to work toward greater efficiencies with plans to migrate to Software-as-a-Service offerings or consolidate further internally.

VA will continue to eliminate duplication and drive down costs while improving service for commodity information technology. A report on the commodity services, proposed for migration to shared services, will be provided by June 30, 2012.

Appendix VI: GAO Contact and Staff Acknowledgments

GAO Contact

David A. Powner, (202) 512-9286 or pownerd@gao.gov

Staff Acknowledgments

In addition to the contact named above, individuals making contributions to this report included Colleen Phillips (Assistant Director), Cortland Bradford, Rebecca Eyler, Kathleen S. Lovett, and Jessica Waselkow.

Related GAO Products

Best Practices in Information Technology Acquisition	*Information Technology: Critical Factors Underlying Successful Major Acquisitions.* GAO-12-7. Washington, D.C.: October 21, 2011.
CIO Responsibilities	*Federal Chief Information Officers: Opportunities Exist to Improve Role in Information Technology Management.* GAO-11-634. Washington, D.C.: September 15, 2011. *Information Technology: VA Has Taken Important Steps to Centralize Control of Its Resources, but Effectiveness Depends on Additional Planned Actions.* GAO-08-449T. Washington, D.C.: February 13, 2008.
Cloud Computing	*Information Security: Additional Guidance Needed to Address Cloud Computing Concerns.* GAO-12-130T. Washington, D.C.: October 6, 2011. *Information Security: Governmentwide Guidance Needed to Assist Agencies in Implementing Cloud Computing.* GAO-10-855T. Washington, D.C.: July 1, 2010. *Information Security: Federal Guidance Needed to Address Control Issues with Implementing Cloud Computing.* GAO-10-513. Washington, D.C.: May 27, 2010.
Data Center Consolidation	*Data Center Consolidation: Agencies Need to Complete Inventories and Plans to Achieve Expected Savings.* GAO-11-565. Washington, D.C.: July 19, 2011.
Duplication and Overlap	*Follow-up on 2011 Report: Status of Actions Taken to Reduce Duplication, Overlap, and Fragmentation, Save Tax Dollars, and Enhance Revenue.* GAO-12-453SP. Washington, D.C.: February 28, 2012. *2012 Annual Report: Opportunities to Reduce Duplication, Overlap and Fragmentation, Achieve Savings, and Enhance Revenue.* GAO-12-342SP. Washington, D.C.: February 28, 2012. *Information Technology: Departments of Defense and Energy Need to Address Potentially Duplicative Investments.* GAO-12-241. Washington, D.C.: February 17, 2012.

Related GAO Products

Information Technology: Potentially Duplicative Investments Exist at the Departments of Defense and Energy. GAO-12-462T. Washington, D.C.: February 17, 2012.

Opportunities to Reduce Potential Duplication in Government Programs, Save Tax Dollars, and Enhance Revenue. GAO-11-318SP. Washington, D.C.: March 1, 2011.

Information Technology Investment Management

Investment Management: IRS Has a Strong Oversight Process but Needs to Improve How It Continues Funding Ongoing Investments. GAO-11-587. Washington, D.C.: July 20, 2011.

Information Technology: Investment Oversight and Management Have Improved but Continued Attention Is Needed. GAO-11-454T. Washington, D.C.: March 17, 2011.

Information Technology: Treasury Needs to Strengthen Its Investment Board Operations and Oversight. GAO-07-865. Washington, D.C.: July 23, 2007.

Information Technology: DHS Needs to Fully Define and Implement Policies and Procedures for Effectively Managing Investments. GAO-07-424. Washington, D.C.: April 27, 2007.

Investment Review and Office of Management and Budget Oversight

Information Technology: OMB Needs to Improve Its Guidance on IT Investments. GAO-11-826. Washington, D.C.: September 29, 2011.

Information Technology: Management and Oversight of Projects Totaling Billions of Dollars Need Attention. GAO-09-624T. Washington, D.C.: April 28, 2009.

Information Technology: OMB and Agencies Need to Improve Planning, Management, and Oversight of Projects Totaling Billions of Dollars. GAO-08-1051T. Washington, D.C.: July 31, 2008.

Information Technology: Further Improvements Needed to Identify and Oversee Poorly Planned and Performing Projects. GAO-07-1211T. Washington, D.C.: September 20, 2007.

Related GAO Products

Information Technology: Improvements Needed to More Accurately Identify and Better Oversee Risky Projects Totaling Billions of Dollars. GAO-06-1099T. Washington, D.C.: September 7, 2006.

Information Technology: Agencies and OMB Should Strengthen Processes for Identifying and Overseeing High Risk Projects. GAO-06-647. Washington, D.C.: June 15, 2006.

Information Technology Dashboard	*IT Dashboard: Accuracy Has Improved, and Additional Efforts Are Under Way To Better Inform Decision Making.* GAO-12-210. Washington, D.C.: November 7, 2011.

Information Technology: Continued Attention Needed to Accurately Report Federal Spending and Improve Management. GAO-11-831T. Washington, D.C.: July 14, 2011.

Information Technology: Continued Improvements in Investment Oversight and Management Can Yield Billions in Savings. GAO-11-511T. Washington, D.C.: April 12, 2011.

Information Technology: OMB Has Made Improvements to Its Dashboard, but Further Work Is Needed by Agencies and OMB to Ensure Data Accuracy. GAO-11-262. Washington, D.C.: March 15, 2011.

Information Technology: OMB's Dashboard Has Increased Transparency and Oversight, but Improvements Needed. GAO-10-701. Washington, D.C.: July 16, 2010.

Information Technology: Federal Agencies Need to Strengthen Investment Board Oversight of Poorly Planned and Performing Projects. GAO-09-566. Washington, D.C.: June 30, 2009.

Related GAO Products

Information Technology Acquisition Professionals

Federal Housing Administration: Improvements Needed in Risk Assessment and Human Capital Management. GAO-12-15. Washington, D.C.: November 7, 2011.

Information Technology: HUD's Expenditure Plan Satisfies Statutory Conditions and Implementation of Management Controls Is Under Way. GAO-11-762. Washington, D.C.: September 7, 2011.

Information Technology: FBI Has Largely Staffed Key Modernization Program, but Strategic Approach to Managing Program's Human Capital Is Needed. GAO-07-19. Washington, D.C.: October 16, 2006.

(311253)

GAO's Mission	The Government Accountability Office, the audit, evaluation, and investigative arm of Congress, exists to support Congress in meeting its constitutional responsibilities and to help improve the performance and accountability of the federal government for the American people. GAO examines the use of public funds; evaluates federal programs and policies; and provides analyses, recommendations, and other assistance to help Congress make informed oversight, policy, and funding decisions. GAO's commitment to good government is reflected in its core values of accountability, integrity, and reliability.
Obtaining Copies of GAO Reports and Testimony	The fastest and easiest way to obtain copies of GAO documents at no cost is through GAO's website (www.gao.gov). Each weekday afternoon, GAO posts on its website newly released reports, testimony, and correspondence. To have GAO e-mail you a list of newly posted products, go to www.gao.gov and select "E-mail Updates."
Order by Phone	The price of each GAO publication reflects GAO's actual cost of production and distribution and depends on the number of pages in the publication and whether the publication is printed in color or black and white. Pricing and ordering information is posted on GAO's website, http://www.gao.gov/ordering.htm. Place orders by calling (202) 512-6000, toll free (866) 801-7077, or TDD (202) 512-2537. Orders may be paid for using American Express, Discover Card, MasterCard, Visa, check, or money order. Call for additional information.
Connect with GAO	Connect with GAO on Facebook, Flickr, Twitter, and YouTube. Subscribe to our RSS Feeds or E-mail Updates. Listen to our Podcasts. Visit GAO on the web at www.gao.gov.
To Report Fraud, Waste, and Abuse in Federal Programs	Contact: Website: www.gao.gov/fraudnet/fraudnet.htm E-mail: fraudnet@gao.gov Automated answering system: (800) 424-5454 or (202) 512-7470
Congressional Relations	Katherine Siggerud, Managing Director, siggerudk@gao.gov, (202) 512-4400, U.S. Government Accountability Office, 441 G Street NW, Room 7125, Washington, DC 20548
Public Affairs	Chuck Young, Managing Director, youngc1@gao.gov, (202) 512-4800 U.S. Government Accountability Office, 441 G Street NW, Room 7149 Washington, DC 20548

Please Print on Recycled Paper.

www.ingramcontent.com/pod-product-compliance
Lightning Source LLC
Chambersburg PA
CBHW081625170526
45166CB00009B/3099